A Ballad for the Birds

Mariah Luamanu

ISBN: 978-1-7344147-2-1

Printed and bound in the United States of America

Edited & Published by Mariah Luamanu
Mariahluamanu@gmail.com

Dedication:

For those who soar high,
without feathered wings...

For the ones who find joy,
in life's simple things.

Soaring high before dawn,
with a bright early start...

tweets and chirps fill the air,
melodies from the heart.

God surely knew
that these live little planes...

would be just like lifeblood,
running through the earths
veins.

The birds keep the insects
from making a scene...

they pollinate flowers
and turn forests green.

Building their homes
in the arms of a tree...

awaiting the moment
their babies break free.

The tiniest one,
a bee hummingbird...

only two inches long,
a size quite absurd.

An ostrich receives,
the "largest" award...

even though they don't fly,
they cannot be ignored.

Peacocks have feathers
that reach six feet tall...

Penguins dive deep,
with a loud and clear call.

Ten thousand species,
how can that be?...

flocks fly in formation,
sharing who leads the V.

Brilliant bright colors,
just look at that beak...

their brains no small matter,
many birds even speak.

Dainty and feathered,
the gems of the sky...

wings that do wonders,
'tis a dream just to fly.

Mariah Luamanu lives in Lolo, Montana with her husband and two daughters. She was born and raised in Alaska and has spent most of her life there. She enjoys sharing her poetry about God and His creation. All of her books can be found on Amazon.

Printed in Dunstable, United Kingdom

91222391R00016